Materials that Matter

PLASTICS

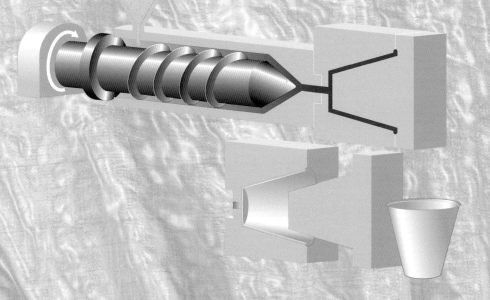

Neil Morris

Published by Amicus
P.O. Box 1329
Mankato, MN 56002

U.S. publication copyright © 2011 Amicus. International copyright reserved in all countries. No part of this book may be reproduced in any form without written permission from the publisher.

Printed in the United States of America,
at Corporate Graphics, North Mankato, Minnesota.

Library of Congress Cataloging-in-Publication Data

Morris, Neil, 1946-
 Plastics / by Neil Morris.
    p. cm. -- (Materials that matter)
 Includes bibliographical references and index.
 Summary: "Discusses plastic as a material, including historical uses, current uses, manufacturing, and recycling"--Provided by publisher.
 ISBN 978-1-60753-068-8 (library binding)
 1. Plastics--Juvenile literature. I. Title.
TA455.P5M665 2011
668.4--dc22

2009051435

Created by Appleseed Editions Ltd.
Designed by Helen James
Edited by Mary-Jane Wilkins
Artwork by Graham Rosewarne
Picture research by Su Alexander

Photograph acknowledgements
Page 4 Ted Horowitz/Corbis; 5 Biosym Technologies, Inc./Science Photo Library; 6 Swim Inc 2, LLC/Corbis; 7 Christopher Jones/Alamy; 8 The Advertising Archive; 9 Bettmann/Corbis; 10 Charles O'Rear/Corbis; 11 How Hwee Young/epa/Corbis; 13 Roger Ressmeyer/Corbis; 14 Craig Lovell/Corbis; 15 Rika/dpa/Corbis; 16 Germany Images David Crossland/Alamy; 17 Phil M Rogers/Alamy; 18 Eddie Linssen/Alamy; 19 Dana Hoff/Beateworks/Corbis; 20 CuboImages SRL/Alamy; 21 Roger Ressmeyer/Corbis; 22 Danita Delimont/Alamy; 23 CarterPhotographic/Alamy; 24 Gari Wyn Williams/Alamy; 25 Chris Ratcliffe/Alamy; 26 Jose Luis Pelaez/Blend Images/Corbis; 27 67Photo/Alamy; 28 Patrick Bennett/Corbis; 29 Directphoto.org/Alamy
Front cover: Stan Gamester/Alamy

DAD0041
32010

9 8 7 6 5 4 3 2 1

Contents

What Are Plastics?	4
The First Plastics	6
A Century of Discoveries	8
The Stages of Making Plastics	10
Shaping Plastics	12
Remarkable Rubber	14
Plastics in Art and Design	16
Plastics Take Over	18
Natural Raw Materials	20
Endlessly Useful	22
Problems and Solutions	24
Recycling Plastics	26
Plastics in the Future	28
Glossary and Web Sites	30
Index	32

What Are Plastics?

Many things are made of plastic, including pens, computer keyboards, light switches, mobile phones, sunglasses, and television sets. There are many different kinds of plastic—some hard, some soft. Most are smooth, but they can have a rough surface, too. The material they are made of is not natural, such as iron ore or wood. Plastic is made from other raw materials.

These colorful plastic shapes are part of a drawing toy called Spirograph. The colors make the pieces more attractive.

Long Chains

Plastics are made from **organic chemicals**. This means that they contain the **element carbon**. Many organic chemicals also contain **hydrogen**, and these are called **hydrocarbons** (because they mix hydrogen and carbon). One of the biggest sources of hydrocarbons is **petroleum**, or **crude oil**.

Chemicals are made up of tiny particles called **atoms**, as all substances are. These particles join together to form **molecules**. We make plastics by joining lots of molecules together in a long chain. These chains of molecules are called **polymers**. We use heat and pressure, and add other substances, to turn organic chemicals into polymers and make plastics.

Making Polyethylene

Scientists make a gas called **ethylene** from petroleum. This gas is made up of molecules of carbon and hydrogen. Scientists discovered that they could make a plastic substance by linking thousands of ethylene molecules together in a long chain. Today they have several ways of doing this, using chemical reactions with other substances.

Molded Material

The word "plastic" comes from an ancient Greek word meaning to mold. We call the material plastic because we can easily mold it into any shape when it is soft. Plastic products are made in all kinds of shapes.

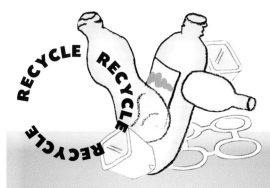

Use It Again and Again

Many types of plastic can be used again and again (see pages 26–27). It is impossible to tell whether a plastic bag or bottle is newly made or recycled. Every time bags and bottles are recycled, the environment benefits. Recycling plastic instead of making new plastic from scratch:

- uses two-thirds less energy;
- uses nearly 90 percent less water;
- emits 60 percent less **carbon dioxide** (CO_2);
- produces two-thirds less other polluting gases.

This makes a plastic called polyethylene. Factories use this to make drink bottles, plastic bags, and many other things. The world's plastic factories make more than 66 million tons (55 million t) of polyethylene every year.

This model shows a chain of molecules that have joined together in the shape of a grid. This is one way in which molecules form a polymer that makes plastic.

The First Plastics

Chemists started making plastics in the 1800s. At first, inventors developed natural substances, such as rubber (see page 14). In 1856, a French inventor mixed animal blood with sawdust, making a material he called *bois durci* (hardened wood). Six years later, the world's first real plastic was made.

Early Mixtures

At the London International Exhibition in 1862, the British chemist Alexander Parkes (1813–90) showed a completely new material. He made it from a mixture of **cellulose**, which comes from the cell walls of plants, and other chemicals. Parkes called it Parkesine. Five years later, his colleague Daniel Spill (1832–87) produced Xylonite, a mixture of cellulose, **camphor**, and **castor oil**. These first plastics were used to make handles for knives and brushes. But they were expensive to make and did not sell well.

Prize-winning Inventor

Billiard balls were once made from **ivory**, which was scarce and expensive. A billiard ball manufacturer offered a $10,000 prize for the inventor of a new material. American printer and inventor John Wesley Hyatt (1837–1920) wanted to win the prize. In 1870, he produced a plastic by heating cellulose, camphor, and alcohol. He called the material **celluloid**, and it was soon used to make buttons, combs, false teeth, and photographic film, as well as billiard balls. Hyatt and his brother built machinery to make celluloid objects, making them easier and cheaper to produce.

This advertisement comes from the 1890s. It tried to show that celluloid was strong and waterproof.

All these objects are made of Bakelite. It was advertised as "the material of a thousand uses."

Whose Invention?

Some plastics with different names are really the same substance, but inventors want to claim them as their own. In a famous court case, Daniel Spill claimed that he discovered celluloid, although he did not make up or use that name. It was a trademark of Hyatt's Celluloid Manufacturing Company. But the American judge ruled that the celluloid process had not been invented by Spill or Hyatt, but by Alexander Parkes.

A New Way of Recycling

In the 1900s, no one knew how to recycle phenol plastics (modern Bakelite). Then in 2005, a Japanese company found a way of recycling the phenol **resin** used in electrical components. They crush the scrap plastic into a fine powder and heat it with water and other chemicals to make new plastic.

New Breakthrough

In 1907, a chemist called Leo Hendrik Baekeland (1863–1944) developed a new plastic. He combined **phenol** (from coal and wood tar) with **formaldehyde** (a gas containing carbon, hydrogen, and oxygen). He called this plastic **Bakelite**. It was a huge success and was used instead of celluloid for many things, especially electrical appliances, including radios and telephones.

A Century of Discoveries

Scientists learned a great deal from early plastics. Manufacturers looked for new materials, and chemists even found some by accident. They could make companies rich. During the 1900s, oil companies found a lot of petroleum, which produced more raw materials for the industry. Space research, medicine, design, and architecture all used plastics.

Accidental Discoveries

The German chemist Hans von Pechmann (1850–1902) discovered polyethylene when he accidentally overheated another substance. A similar mistake led to a different form, called low-**density** polyethylene (LDPE). Chemists in Britain produced this in 1933. Twenty years later, another German chemist, Karl Ziegler (1898–1973), discovered a high-density form (HDPE). This was used to make bottles and other containers, and Ziegler won the Nobel Prize for chemistry for his work. In the 1970s, the first plastic shopping bags were made from LDPE.

Parachutes and Stockings

Wallace Carothers (1896–1937) was an American chemist who worked for the DuPont Company in Delaware. In 1935, Carothers and his team used chemicals in coal, petroleum, air, and water to make a tough plastic that could be stretched into fibers. They called this nylon. Three years later, they launched their first product—a toothbrush with nylon bristles.

A 1940s advertisement for nylon stockings also shows nylon fishing nets, hair bows, and sails.

In 1939, the first nylon stockings created a sensation in San Francisco. Before then, women's stockings were made from silk and

The first fiberglass Corvettes roll off the assembly line in 1953. They were built at the Chevrolet factory in Flint, Michigan.

were very expensive. When nylons reached American stores, they cost a dollar a pair and women bought five million pairs on the first day. This was during World War II, when nylon replaced silk in parachutes and other war equipment. Nylon stockings became more widely available after 1945.

The First Plastic Car

In the mid-1900s, scientists realized that glass could strengthen plastics. This material is called glass-reinforced plastic (or GRP), but people usually call it **fiberglass**. In 1953, the American car maker General Motors made a sports car with a fiberglass body. This was the revolutionary Chevrolet Corvette. Today, fiberglass is used more often for boats.

Plastic Discoveries

1862 Parkesine
1869 Xylonite
1870 Celluloid
1892 Viscose
1907 Bakelite
1929 Polystyrene
1930 Polyester
1933 Melamine, polyvinylchloride (PVC), and polyethylene (LDPE)
1935 Nylon
1939 Polyurethane
1943 **Silicones**
1954 Polypropylene

The Stages of Making Plastics

There are two stages in making plastic products. First, a manufacturer uses chemicals and other raw materials to make the basic plastic material, called resin. This is usually in the form of powder, small granules, or larger pellets. Manufacturers then melt the resin and shape it into products (see pages 12–13).

How Plastic Resin is Made

Plastic resin such as polystyrene is made in a chemical factory. First, bubbles of ethylene gas (from petroleum) are passed through liquid **benzene** (also from oil). The liquid changes into ethylbenzene, which is heated with metallic chemicals to produce styrene. The styrene is put in water and heated with more chemicals, so thousands of styrene molecules join together to form solid polystyrene. Finally, a machine cuts the polystyrene into granules. These are used to make toys, computers, or TV cabinets.

Making Better Plastics

Manufacturers often add substances to the resin, to improve the quality.

These pellets of plastic resin come in a range of colors.

This resin will be turned into plastic containers at a factory in Singapore.

Glass fibers make the plastic stronger or harder. Powdered clay can improve it and make it cheaper. **Plasticizers** are substances that make the material more flexible and easier to shape. **Lubricants** make sure the resin does not stick to molding equipment. Finally, pigments change the color of the plastic, so items can be made in a range of different colors.

Two Types

There are two basic types of plastic. One type, thermosets, do not soften and melt when heated. These include Bakelite and melamine, and they are useful for products which withstand heat, such as pot and pan handles. They can also take a lot of pressure, so are good for tabletops and chairs.

Thermoplastics do melt when they are heated. Polyethylene, for example, melts at a temperature of 221–266°F (105–130°C), so it cannot be used for cookware. Thermoplastics are generally easier and quicker to produce. They make up about 90 percent of the world's plastic production.

More and More Resin

We are making and using more plastic resin all the time. By 2010, we were using almost twice as much plastic resin as we did in 2002.

Shaping Plastics

Engineers shape plastic resins by heating the resin granules, until they are very soft. Then items can be mass produced, so they are all exactly the same shape and size, by molding or casting. Waste material from these processes is recycled at the factory (see pages 26–27).

Casting

Molten plastics are cast by pouring liquid resin into a mold. Thermoplastics are then left to cool, until they set. Thermosets are different, as they are hardened through heating.

Molding

There are several different methods of molding plastics.

Blow molding Air or steam blows up a lump of runny resin, like a balloon. The resin presses against the inside of a mold to make hollow shapes, such as bottles.

Injection molding A turning screw forces the resin into a mold, where it becomes solid and takes its shape.

Rotational molding The mold spins very fast, so that the resin spreads around the inside of the mold.

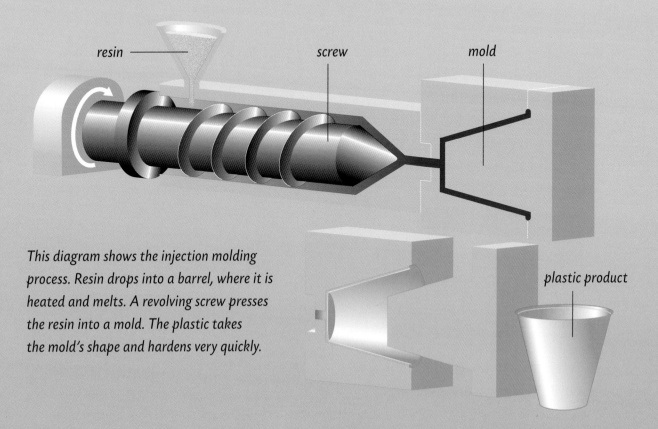

This diagram shows the injection molding process. Resin drops into a barrel, where it is heated and melts. A revolving screw presses the resin into a mold. The plastic takes the mold's shape and hardens very quickly.

Vacuum molding A soft sheet of resin is put over a mold, and the air is sucked out. The sheet takes on the shape of the mold.

Pressing and Coating

Pressing molten resin between pairs of heated rollers makes plastic sheets or even thinner plastic film. Manufacturers can feed paper, aluminum foil, or cloth through the rollers to give them a plastic coating.

Making Plastic Foam

There are lots of different foam plastics. Hard, stiff foam is used to make food trays and cartons, or protective cases for toys and electronic goods. Soft, spongy foam is used for furniture cushions.

Foam resins are made by mixing the resin with chemicals that give off gas, which makes bubbles in the cooling resin. Foam plastics are very light, which makes them useful for packaging and insulation, because the trapped gas bubbles do not pass heat easily and are excellent insulators.

Light and flexible polystyrene foam chips make useful packing material. They are sometimes called packing peanuts.

Saving Oil

Thermoplastics are easy to recycle, because they can be melted down to make new resin. This saves a great deal of a basic raw material—oil. We use about eight percent of the oil we produce every year to make plastics. Every time we recycle a ton of plastic bottles (about 20,000 two-liter bottles) we use about four barrels or 168 gallons (636 L) less oil than if we made the bottles from scratch.

Remarkable Rubber

Some types of plastic are squishy and stretchy. You can pull or squeeze them, and they return to their original shape when you let go. Scientists call these substances **elastomers**. We call some of them rubber, which is the name of a natural substance that comes from the juice of rubber trees.

Today, most rubbery materials and products are made from plastic resin or a mixture of plastic and natural rubber. There is not enough natural rubber to supply the world's needs, so **synthetic** rubber has become a very important material.

Discovering the Eraser

In the early 1700s, French explorers in Peru found Native Americans using the dried and hardened juice of a tree they called "cahuchu." The explorers took samples back to Europe, where they called it "caoutchouc" until 1770. Then the chemist Joseph Priestley (1733–1804) noticed that the material erased or rubbed out pencil marks. He renamed it rubber, and that has been its name ever since.

Stronger Tires

In 1839, U.S. inventor Charles Goodyear (1800–60) found a new way to use natural rubber. He strengthened the sticky material by heating it and adding **sulphur**. This made the rubber firmer, whether it was hot or

A villager in Thailand harvests white latex from a cut in a rubber tree.

This worker is checking a tire at a factory in Brazil.

cold. The new material was called vulcanized rubber, after the Roman god of fire, Vulcan. At the end of the nineteenth century, it was used to make tires. Goodyear tires, first produced in the United States in 1898 and still sold around the world today, were named in honor of the inventor.

Making Rubber

Today, more than 90 percent of all natural rubber comes from rubber tree plantations in Indonesia, Malaysia, and Sri Lanka. Tree tappers cut the bark of the trees and collect the juice (or **latex**) that runs out, then take it to a factory. Acid is used to separate the rubber from water in the latex. Rollers then press the rubber to get rid of more water.

Synthetic rubbers are made in a similar way to other plastics. Chemical substances, mainly from petroleum, are combined, heated, and treated. Both natural and synthetic rubbers are molded into products.

Recycling Rubber

Two-thirds of the world's rubber is used to make tires for vehicles. When they wear out, tires can be reused for other purposes, such as bumpers for boats. Rubber is a thermoset material (see page 11), so it cannot be melted down and reused. But old rubber goods, especially tires, can be ground into granules. These can be mixed with new synthetic rubber and used to make mats, shoe soles, or road surfaces. Recycled rubber is not good enough to make new tires.

Plastics in Art and Design

Plastics are very practical materials, because they come in all shapes and colors. This makes them attractive to artists and designers. Plastic designs are clean and shiny, and this look is popular with interior designers and homeowners.

There were 90 professional climbers and 120 installers who helped Christo and Jeanne-Claude wrap Germany's parliament building.

Plastic Cover-up

Some sculptors design objects in plastic rather than traditional stone or metals such as bronze. One famous pair of American artists, Christo and Jeanne-Claude, use plastic fabrics to create works of art by covering things up. Bulgarian-born Christo Javacheff and French-born Jeanne-Claude Denat de Guillebon (both born in 1935) have covered entire landscapes and buildings in plastic. In 1965, they covered 969,000 square feet (90,000 sq m) of coast near Sydney, Australia. In 1995, they used 108,000 square feet (100,000 sq m) of polypropylene fabric to cover the Reichstag

Recycled Sculpture

The Canadian sculptor Aurora Robson (born in 1972) uses all kinds of materials in her art. They include discarded plastic bottles, which she turns into fascinating sculptures.

These used bottles have been painted in bright colors and recycled as lights.

building in Berlin, Germany. All the plastic materials were later recycled. Then in 2005, they put up 7,500 vinyl gates with nylon panels in Central Park, New York.

Fun Furniture

Interior designers are creating more and more plastic furniture. Plastics make practical kitchen worktops and bathroom surfaces. They last well, too. Plastic is an important material in the modern style known as **New Design**. French designer Philippe Starck (born in 1949) has created plastic shapes for chairs which are now style classics. Many are made of a **polycarbonate** material in a range of colors.

Keeps Out the Weather

Plastic materials are strong and weatherproof, but architects did not start using them on the outside of buildings until the 1970s. The stadium built for the Munich Olympic Games in 1972 is covered by a canopy roof made of 805,140 square feet (74,800 sq m) of acrylic glass. This is the plastic material often called perspex or plexiglass. The stadium roof is supported by steel cables attached to 58 masts, and it also covers an Olympic hall and swimming pool. The Millennium Dome (The O_2) in London also has a canopy roof, made of 108,000 square feet (10,000 sq m) of fiberglass fabric coated with a nonstick plastic. The canopy is held up by 43 miles (69 km) of steel cable attached to 12 towers.

Plastics Take Over

Many containers are now made of plastic rather than glass. Plastic has replaced glass in eye glasses and in aircraft windows because it is lighter and less **brittle**. It is used instead of paper in many forms of packaging and sometimes instead of metals and wood. This shows how versatile plastic is.

Some football stadiums have plastic seats in the colors of the home team, as here at FC Barcelona's Camp Nou stadium.

Lighter than Metals

A plastic part in a car is about one-fifth as heavy as the same part made of steel. It can be molded to exactly the right size and shape very easily, which makes the part cheaper. Bumpers, hubcaps, and other car accessories were once made of steel. Today, they are all plastic, making them lighter, cheaper, and easier to replace, if they are damaged. Making cars lighter means they use less fuel, too. In homes, modern water pipes are made of plastic instead of lead,

Nonstick Coating

The nonstick plastic called polytetrafluoroethylene (PTFE) is hard and tough, and it resists heat and electricity. PTFE melts at a temperature of 621°F (327°C), about the same as lead. Most important of all, this plastic is extremely slippery, so it can coat metals and make them nonstick. This is particularly useful for frying pans and other cookware. A nonstick plastic coating can be added to aluminum, cast iron, or steel.

It's easy to fry eggs in a nonstick pan.

copper, iron, or steel. Plastic pipes are easier to cut and join, and they do not wear away.

PVC and Pollution

In some modern buildings, the doors and window frames are made from PVC rather than wood. Some people prefer plastic, because it does not need painting, and it lasts a long time. Others say that plastic never looks as attractive as wood. The environmental group Greenpeace says that PVC is harmful to the environment. It claims that PVC is one of the world's largest sources of poisonous **dioxins**. PVC manufacturers deny many of Greenpeace's claims. They quote reports that dioxin levels due to PVC manufacture are "very small."

Recycling PVC

PVC can be ground into pellets, reheated, and molded into new products. It can be recycled up to ten times. One problem with recycling PVC is that it is often contaminated by other materials. Separating and sorting waste PVC is expensive. New methods of recycling are being developed to deal with this problem.

Natural Raw Materials

The raw materials for making plastics come mainly from oil, and supplies of oil are running out. Burning and refining oil is harmful to the environment because it causes pollution. Recently scientists discovered a new way to produce plastic from plants.

These rows of corn plants could be used to make a form of bioplastic.

New Source

The material made from plants is called bioplastic or organic plastic. One of the most common is polylactide (PLA), which comes from a substance in sugar cane or corn. Other bioplastics can be made from potatoes, **cassava**, and **soybeans**. Researchers are looking for ways of turning more plants into plastic.

Plastic versus Food

Environmentalists want to use less oil, but there are problems with bioplastics. If farmers grow crops for making plastic rather than food, this would be a disaster for people who can't grow food or who don't have enough money to buy it. Plants are already grown for biofuels rather than food, and bioplastics might make the situation worse.

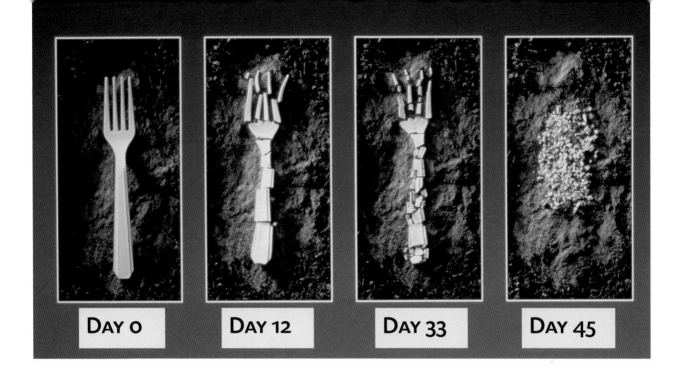

| Day 0 | Day 12 | Day 33 | Day 45 |

Improving Plants?

Some companies are trying to improve plants for plastics by changing their genes. All plants (and animals) have biological sets of instructions within their cells. These coded instructions are carried by genes, which pass on the code to the next generation. Scientists have learned how to alter a plant's genes. This is called genetic modification (GM), and many environmentalists are against it. They say that GM crops are unnatural and may turn out to be dangerous to both the environment and human health. They believe that GM crops could spread, wiping out natural plants and reducing the world's range of plants and animals.

Making Compost

Food waste can be recycled to make compost, which is a mixture of decayed organic matter that gardeners use to make soil richer. Bioplastics can be recycled

This bioplastic fork was made from a soft corn-based polylactide. The photographs show how it decomposed over 45 days.

in the same way, and so they are called compostable. Plastic trays made from sugarcane and PLA garbage bags can break down in three months. Hard utensils made from cornstarch may take up to 18 months.

Recycling Issues

One of the problems with bioplastics is that they cannot be recycled together with plastics from oil. The two different types need to be separated, and bioplastics need their own code number (see page 27). Some experts have suggested that another way around this problem is to have different shapes for bioplastic bottles.

Endlessly Useful

Plastics have many different uses. One group of plastics called acrylics is made from acrylic acid—a colorless, smelly liquid. Acrylics can be turned into fibers, plastic glass, paints, and many other materials.

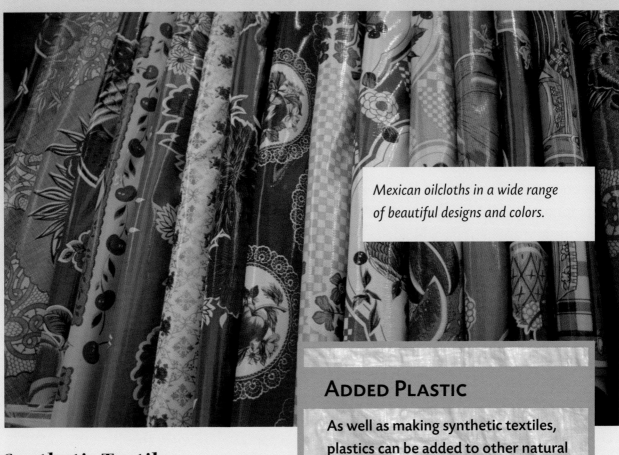

Mexican oilcloths in a wide range of beautiful designs and colors.

Synthetic Textiles

Machines with tiny holes can turn acrylic resin into long, thin fibers. These can be used to make synthetic textiles for hats, rugs, sweaters, and upholstery. Acrylic textiles are light, soft and warm, and can be dyed any color. They are sometimes used instead of wool, and they are generally much cheaper.

Added Plastic

As well as making synthetic textiles, plastics can be added to other natural materials. Cotton cloth can be passed through runny plastic resin so the plastic soaks into the natural fibers of the cloth. Another way to coat natural fibers with plastic is to press a warm sheet of plastic (such as PVC) onto the cotton cloth. The material is called oilcloth and is used for wipeable, waterproof tablecloths.

Acrylics can look very similar to oil paints.

See-through Materials

Acrylic glass (often called perspex or plexiglass, see page 17) was developed during the 1930s and was used in World War II (1939–45) for aircraft windshields. English eye specialist Harold Ridley (1906–2001) had to remove pieces of perspex from the eyes of wounded pilots, which gave him the idea of using acrylic glass for artificial lenses. In 1950, he implanted his first lens, and patients with eye problems such as **cataracts** are still given acrylic lenses today. Each tiny, precision-made lens measures just 0.24 inches (6 mm) across.

Modern Paints

Acrylic paints were first made in the 1950s, by mixing a pigment (or colored substance) with a runny acrylic resin and then adding water. The paints are easy to use and dry quickly, and artists can make them look like oils or watercolors. Once dry, the paintings are waterproof, and they last better than other paints. Another advantage of acrylics is that you can add things to the wet surface, such as sand, or mark the dry surface with charcoal, pastel, or pencil.

STICKY STUFF

Acrylic resins can also make glues and elastic materials, which seal cracks in walls—yet another use for plastic.

23

Problems and Solutions

We use about 20 times more plastic today than 50 years ago. One of the big problems with plastics is that they last for a very long time but are easy to throw away. Plastic litters our streets, parks, countryside, and beaches. We could all work harder to solve this problem.

Breaking Down Plastic

Natural materials, such as leather or wood, biodegrade or break down if they are dumped into a landfill. They decay relatively quickly, as **bacteria** break them down, but most plastics do not decay easily. They may last for hundreds or even thousands of years. In recent years, manufacturers have produced biodegradable plastics. They add substances to the resin that make it decay more quickly. The problem is that when they biodegrade, plastics are broken down into carbon dioxide and when they rot in landfills, they give off **methane**. Both are greenhouse gases that contribute to the problem of **global warming**.

Plastic litter is difficult to clean up and can hang around for many years.

If shoppers remember to take bags with them, they save plastic and help the environment.

Pollution at Sea

Plastic trash is not just a problem on land. Environmental experts say there is a huge marine trash pile in the North Pacific Ocean, halfway between East Asia and North America. They estimate that this pollution may cover millions of square miles (sq km) and contain more than 3.3 million tons (3 million t) of plastic. Most of the trash comes from land and a small amount from ships at sea.

Cleaning Up the Ocean

In 2008, a group of environmentalists founded the Environmental Cleanup Coalition. Their aim is to tell people about the problems of marine pollution. They want to clean up the world's oceans and have started by sending specially designed ships to catch plastic trash in nets. They then recycle the waste.

The Three Rs

We could all waste less plastic by following the three Rs. The Rs stand for:
- reduce
- reuse
- recycle.

Reducing means cutting down on waste by using less in the first place. The most obvious way to cut down on plastic waste is to use less packaging. Fresh food is often wrapped in plastic and sold on plastic trays, but we could easily buy fruit, vegetables, and other food items without packaging.

We could reuse more plastic, for example, by reusing plastic bags when we go shopping. Finally, instead of throwing away plastic containers in an ordinary trash can, we can put them in a recycling bin. Then, instead of being dumped, the old plastic will be reused to make new containers.

Recycling Plastics

We recycle plastics by melting them and reusing the resin to make new products. The number of plastic bottles and other containers we recycle is growing every year. The biggest difficulty is that waste plastic has to be sorted and separated, as different kinds cannot be recycled together.

Biodegradable plastics also have to be sorted, because if they are mixed with other plastics, the material is not recyclable. The plastics industry has a series of code numbers, so that recyclers can tell one plastic from another.

Plastic Recycling Code

In 1988, the Society for the Plastics Industry (SPI) set up a coding system. They numbered the six most common types of plastic. The numbers appear on the bottom of containers, so that recyclers know what type of plastic they are made from. Take a look at the bottoms of some plastic bottles, then check the code and see what type of plastic they are made from.

The Recycling Cycle

Garbage companies collect plastic waste from homes or neighborhood recycling bins. This is called postconsumer waste. They take the plastic to recycling centers to be sorted, often into different colors, and separated from other materials, such as metals. Each type of plastic is then crushed and pressed into bales. Plastic manufacturing plants treat the crushed plastic further and reduce it to small fragments or flakes. Manufacturers use the plastic flakes to make new products.

Many households keep a special trash bin for collecting plastic containers for recycling.

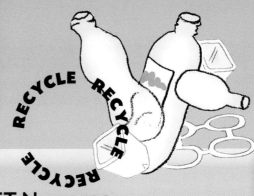

PET Numbers

The amount of postconsumer PET (plastic number 1) bottles collected for recycling in the U.S. was 1.452 billion pounds (658.4 million kg) in 2008. This represented 27 percent of the 5.366 billion pounds (2.4 billion kg) of PET plastic sold that year.

These huge stacks are made up of plastic bottles. The recycling plant will grind them down to small flakes or chips.

What the Numbers Mean

	Type of plastic	Abbreviation	Uses
1	Polyethylene terephthalate	PET or PETE	Soft drink bottles, food trays, carpets
2	High-density polyethylene	HDPE	Bottles, recycling bins, playground equipment
3	Polyvinyl chloride	PVC or V	Pipes, fencing, nonfood bottles
4	Low-density polyethylene	LDPE	Grocery bags, soda can rings, containers
5	Polypropylene	PP	Car parts, food containers, industrial fibers
6	Polystyrene	PS	Office accessories, cafeteria trays, utensils, toys, video cassettes
7	Other plastics, (acrylic, fiberglass, nylon)		

Plastics in the Future

We could call the second half of the twentieth century the age of plastic, when plastics replaced many other materials. Today, people are much more concerned about the environment than they were 50 years ago. They are looking for new sources of plastics, but they also want to see much more of this amazing material reused and recycled.

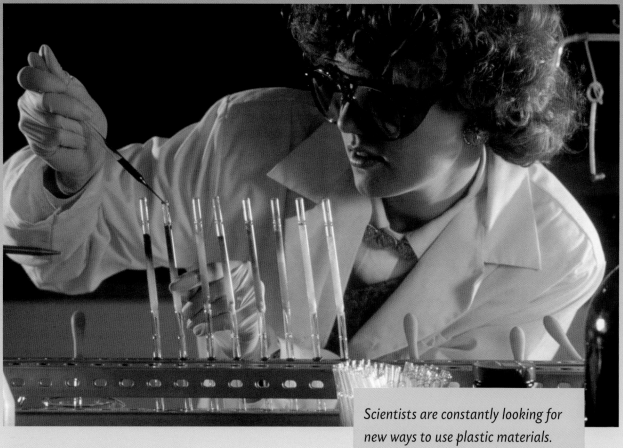

Scientists are constantly looking for new ways to use plastic materials.

Integrated Circuits

Researchers believe that future plastics could transform electrical goods. They have been working for years on developing resins that can carry an electric current. One of the plastics that can do this is called polythiophene (or PT for short), and this is proving to be very successful. Scientists hope that this material could replace **silicon** for making integrated circuits. It could also be used for computer screens that could roll or fold up like a newspaper.

New Sources

Just as the energy industry is looking for renewable sources to replace fossil fuels, the plastics industry will continue to look beyond oil for its raw materials. We already have biodegradable plastics (see page 24) and bioplastics (see page 20). Scientists hope to find more new sources for plastic resins in the future.

Designer Philippe Starck (see page 17) is famous for his colorful plastic chairs. Will more plastic household items be designed in the future?

Artificial Blood

British scientists have created artificial blood from plastic molecules. The tiny particles contain iron, just as natural blood does, and can carry oxygen. One advantage artificial blood has over natural blood, donated by humans, is that the new blood can be stored for much longer. Doctors could store it as a thick paste and then dissolve it in water, before giving patients a blood transfusion.

Better Breakdowns

Tiny organisms called bacteria help materials to biodegrade or decay. Perhaps in the future, bacteria will find new ways to attack plastic. Or scientists might find ways to alter bacteria so that they can do this. In 1975, Japanese scientists found bacteria in ponds of waste water from a nylon factory. They discovered that these bacteria were able to digest certain kinds of nylon. Then in 2008, a 16-year-old Canadian student discovered bacteria that can break down plastic bags.

Color Changes

The scientists who developed the first plastics never dreamed that their materials might change color on their own. But today's researchers talk about intelligent plastic, which can react to changes in light and temperature. Perhaps we will be able to develop new acrylic paints that can change color when different amounts of light shine on them. We could change the color of a room just by turning on a light.

Glossary

atom The smallest part of an element that can take part in a chemical reaction.
bacteria Very tiny organisms that can cause decay.
Bakelite An early kind of plastic.
benzene A colorless liquid which comes from oil.
brittle Hard and likely to break.
camphor A white, strong-smelling substance from the camphor tree.
carbon A chemical element which all living things contain.
carbon dioxide (CO_2) A greenhouse gas given off when we burn fossil fuels, such as coal, oil, and gas.
cassava A tropical plant with edible roots.
castor oil Oil which comes from the seeds of the castor oil plant.
cataract An eye disease where the lens of the eye becomes clouded.
celluloid A transparent kind of plastic which burns.
cellulose A substance that makes up the cell walls of plants.
crude oil Oil (or petroleum) as it is found naturally underground.
density The amount of matter in a unit of a substance; high-density substances have more matter than low-density substances.
dioxin A poison produced when some chemicals are manufactured.
elastomer A material that can be stretched yet returns to its original shape.

element A substance that cannot be separated into a simpler form.
ethylene A colorless gas which comes from oil.
fiberglass A kind of plastic that is reinforced by glass.
formaldehyde A colorless, strong-smelling gas.
global warming Heating up of the Earth's atmosphere and surface, especially caused by pollution from burning fossil fuels.
hydrocarbon A chemical compound containing hydrogen and carbon.
hydrogen A light, colorless gas that combines with oxygen to make water.
ivory The creamy-white bone that forms the tusks of elephants.
latex A milky liquid produced by some trees that is used to make rubber.
lubricant A substance that can make things smooth and slippery.
methane A gas that burns and is the main element in natural gas.
molecule A tiny particle of a substance that is made up of atoms held together by chemical forces.
New Design A modern style or design movement that uses many new materials.
organic chemical A substance that contains the element carbon.
petroleum Crude oil.
phenol A poisonous substance in coal and wood tar; also called carbolic acid.

plasticizer A substance that can make another substance soft and flexible.

polycarbonate A strong kind of plastic that is easily molded.

polymer A substance formed by a chain of molecules.

resin Basic plastic material before it is molded or made into a product.

silicon An element in sand and many minerals; it is used to make integrated circuits.

silicone A plastic material made from silicon.

soybeans Seeds of the soybean plant that are rich in oil.

sulphur A yellow, nonmetallic chemical element.

synthetic Made artificially rather than from natural mterials.

Web Sites

Extensive learning center on plastics by the American Chemistry Council.
http://www.americanchemistry.com/s_plastics/sec_learning.asp?CID=1102&DID=4256

Everything you ever wanted to know about rubber.
http:www.bouncing-balls.com

All about the history of plastics by the Plastics Historical Society.
http:www.plastiquarian.com

History, production, uses and recycling of plastics by the Plastics Industry Trade Assocation.
http://www.plasticsindustry.org/aboutplastics/

Easy-to-understand explanations of how stuff works, including many pictures and diagrams about plastics.
http://science.howstuffworks.com/plastic.htm

Index

acrylics 17, 22, 23, 27, 29
art and design 16, 17, 23

bacteria 24, 29
Bakelite 7, 9, 11
bioplastics 20, 21, 29
blood 6, 29
boats 9, 15
buildings 16, 17, 19

carbon 4, 7
carbon dioxide 5, 24
cars 9, 18, 27
casting 12, 13
castor oil 6
celluloid 6, 7, 9
cellulose 6
chemical reactions 4
compost 21
cookware 11, 19
corn 20, 21

dioxins 19

elastomers 14
environment 5, 19, 20, 21, 25, 28
ethylene 4, 10
eye glasses 18

factories 5, 11, 12, 15
fiberglass 9, 17, 27
formaldehyde 7
furniture 17, 29

genetic modification 21
glass 9, 10, 17, 18, 22
global warming 24
greenhouse gases 24

hydrocarbons 4
hydrogen 4, 7
HDPE 8, 27

latex 14, 15
LDPE 8, 9, 27
litter 24, 25
lubricants 11

methane 24
molding 11, 12, 13, 15, 18, 19
molecules 4, 5

nylon 8, 9, 17, 27 29

oil 4, 8, 10, 13, 20, 21, 29

packaging 13, 18, 25
paper 18
parachutes 9
perspex 17, 23
petroleum 4, 8, 10, 15
phenol 7
pipes 18, 19, 27
plants 20, 21
plastic bags 5, 8, 21, 25, 27, 29
plastic bottles 5, 12, 13, 17, 21, 26, 27

plastic foam 13
plasticizers 11
plexiglass 17, 23
pollution 19, 20, 25
polycarbonate 17
polyester 9
polyethylene 4, 5, 8, 11, 27
polymers 4, 5
polystyrene 9, 10, 13, 27
PTFE 19
PVC 9, 19, 22

raw materials 4, 10, 13, 20, 29
recycling 5, 7, 13, 15, 17, 19, 21, 25, 26, 27, 28
resin 7, 10, 11, 12, 13, 14, 22, 24, 26, 28, 29
rubber 6, 14, 15

sculpture 16, 17
silicones 9

textiles 22
thermoplastic 11, 12, 13
thermoset plastic 11, 12, 15
tires 15

vinyl 17
viscose 9

windows 18, 19
wood 6, 18, 19, 24

32